The Use of a Wall Mirror

The Use of a Wall Mirror

Adriaan Morriën

A Selection from *Mothers and Sons* (1962)
And *The Use of a Wall Mirror* (1968)
Translated from the Dutch by

Ria Leigh-Loohuizen

twowindows press: san francisco

The original Dutch edition of *The Use of a Wall Mirror* was published by Beizge Bij in Amsterdam.

GIRL

She bares her breast like a wound,
as if it hurts to be naked,
like a child showing a butterfly in its hand,
afraid that it will fly away.

All the same her body is nowhere as closed
as at her nipples which, frightened
of shame, would like to flee
like butterflies from a child's hand.

Her hands don't know the luxury
of a loved woman who
lets her fingers rest in her lap because they are
stroked tired and her lips kissed silent.

She gets confused in cold and heat,
doesn't feel the security of the light
and the relationship with her shadow.
Her womb is turned in upon itself and afraid.

palm

I stick my fingers in between
your fingers, into a
palm with separate roots
because I don't yet dare to reach
your breasts in the promised land.

Because I don't send out kisses yet
to prepare a residence,
a shelter, sleeping-quarters
for all my contradictory
fire.

Because I don't want to be tired yet
and burnt-out,
satisfied as a sleeper who
lets his knee rest next to your knee,
while his hand frees itself from your hand.

understanding

She offered me a place in her lap.
I made myself comfortable in her,
stretched myself like somebody who has slept,
and played in the depths of her body.

She had bared herself for my coming,
expressed to me her deeper contentment
and told me that I was announced to her
and commissioned by her heart.

We were lying fondly tangled:
I took her breasts under my supervision,
she blushed in my blood and to my rod
she showed herself a true mother.

I was a father to her sheath.
I let her want for nothing
and supplied her with assistance, provisions and amusement
in abundance. And, touched, she paid me back.

shipwreck

Her breasts stormed out in front of her,
nippled cyclones,
her lap foamed like a sea
and I was shipwrecked on her coast.

I walked ashore on her and everything was unfamiliar:
the strangers who gave me hot rum to drink,
the small wooden room with the view of the harbor,
the sea which had lain itself to rest,
lighthouse, semaphore and even a fort
which wrapped itself in silence,
the silence after a last shot.

This is my father, she laughed.
A grey man looked at me piercingly.

And this my mother.
A much too decided woman
asked the ins and outs of me,
then mended my clothes, which were in rags,
and put me near the fire to dry.

But later we fled.

In the dunes she knew the places where you could descend
to the depth of each other's body.
Space had a big sense of its own dignity.
I kissed her in places where
I'd never before been able
to kiss a woman washed ashore
because I had never wanted to bend down:
between her panties and her stocking,
the pathway to lap and ass
so angelically sweet
that it makes me just as mystical
as at my mother's apron
when I still reached to heaven with my forehead
in smells of urine
elastic
and starch.

And therefore she laughed at me
as if I were the child and not she,
but laid down willingly
under my lips
against my skin.
Between my teeth I carried her home.

Rain

While it's raining now I'm writing
perhaps it won't be raining while you're reading.

And even if it should be raining while you're reading,
then that rain is not comparable to the rain
that I'm describing and have described when you're reading.

This is nonsense and of no importance
but here begins a notion of what is
experienced by us as a mutual misunderstanding.

It's raining dialogue between sky and land.
It's raining kisses, sweat, sleep and meanings
we ascertain by way of weather-reports.

It's raining suspicion: I see it in the gleaming
of the asphalt and the drops on your cheek.
It's raining contingencies and incomprehensible imaginings:
your fingers are trembling and your eyes make tears.

Death and Eternity

Eternity lies behind us:
alive we increase the distance between her and us.
She is the ship, the sky and the sea,
we are the land, the handkerchief and the eye.
The ship gets smaller, sea and sky drain empty,
the land withers, all its rivers fall dry.

Death is the missing, a fullness which disappeared,
but never completely as long as we ourselves still are.
A storm wind full of invisible figures,
but which I perceive as I feel the wind on my leg,
with the hairs on the nakedness of my leg,
the resistance which doesn't keep us from falling.

occupation

I am too old to die:
I have let my time go by.

Once bullets flew through the street:
I did not bare my chest.

Once there was hunger and illness:
I stayed in bed to eat.

Once it was a burden to live:
I have evaded death.

an odd one

He is wiser than all of them
and almost so wise
that he doesn't know a thing.

He feels more at home in the language
than a race of talkers
and can almost be silent.

He lives more intensely than a drunk
but is not afraid
to stay sober.

He is as healthy as a fish
but has no fear
of dying.

speech exercise

Let us forget everything
that has been spoken between us
and speechless as animals
love each other.

To me the light in your eyes,
the silent territories
of shoulders and bosom are sufficient.
To me the saliva is a help.

Let us be silent
glad with a silence which learned to speak,
pleased by speech-defects
which one can taste and smell.

Let us live like animals
which fit narrowly into each other:
language a betrayal to silence
silence a greater embrace.

my parents' house

Language was used sparingly:
caresses had to be stolen.
When my mother stroked me she was ashamed
and I was ashamed —our shames were still separate.

I loved the scissors she cut with,
her thimble, the thread she moistened,
the whirring of the sewing-machine
and the silence when the machine was still.

She slept in a matchless bed
and when my father got up for work
I took his place in her warmth.

I touched her as another child
touches the wall it sleeps against
with life-size fingertips
and without the wall noticing.

the use of a wall mirror

Alone at home, naked on my bed,
I hold the mirror from my parents' room
just above myself and support it with my left hand,
while I masturbate with my right.

I look at the reflection of my white skin,
my thin arms, long legs,
and my tanned face, of a boy still, with
glasses.

Only halfway in love with myself
I try to recognize myself
as reflection and rival—
I too would rather be lying in the arms of a girl.

My eyes look at me: I am not ashamed.
There are no grounds to indulge in the sight of myself.

I listen to the silence in the house,
the only one I trust.

Words originate, as in somebody who flees in fear of death,
without connection and unforgettable.

The trouble it costs to hold the mirror still
and not to slacken
the movement of my right hand: a regularity
upon which I have to keep a strict hand,
trained on other afternoons alone in the house,
or before going to sleep
while I thought of a girl,
or rather a dozen girls
whom I married with lightning speed.

A pianist doesn't just drop out of the sky either.

Return

All revolutions do not wipe out the death
of my mother.
And all progress doesn't prevent
my having hated my father.

Still I always take, too late from school,
the step to the kitchen-door
and lift the latch, afraid
that again the door will crack too loudly.

Everything is possible: my father is silent
or reads me a lecture with severe eyes;
my mother smiles or looks
at her plate with silent eyes.

I draw up to the table
and reach, right through the past,
with a hand become childlike again,
for the steaming bowl of potatoes.

Engagement

Every evening
when she goes to her fiance
and gets off the train
—the train takes twenty minutes,
twenty minutes which go by unnoticed and yet slowly—
she has to pass the men's toilets.

The stink of the urine
only makes her more amorous now
and high-pitched
she opens her eye
at the lucky diversity of the world
in which she has found her way.

moon landscape

Full moon:
a pregnancy which has made itself independent
in phases and deceit.

You think: she's at the point of bursting
and an army of children—
completed girls and fierce boys
stumbling over navel strings,
struggling with amniotic fluid, placenta—
occupies the earth and the dominion
of parents and parents-in-law is over.

But after an hour
the labor pains appear premature.

The nightly life goes its prescribed way.
The urge to vomit has changed
into the need to couple again
and fury has cooled off to beauty.

A boy helps a girl out of her wedding dress
and nestles next to her in adoration of the sun.

one swallow doesn't make a summer

The forest crackled with misfortune

ants barred my way
beetles undermined my shoes
a quail nestled itself in my ear
a woodpecker tapped my chest
and nightingales jugged in my groin

But fortunately, when I reached into
my own bosom
I pulled out three swallows

Together they made summer

REMEMBERANCE

Our children forget the dead
whom we loved
and whom we still commemorate,
suddenly, in strange company,
or walking home deep in the night,
or at the end of the most ordinary day.

The vein in this pebble
resembles the vein in your eye.

There was something transparent in you
that I'll never understand
but always admire.

I feel the wind on my hand
and think of you who woke with a headache,
nevertheless washed yourself cheerfully,
dirtied your white fingers at once.

I smell the struck match,
the smoke which penetrates the chinks of the stove.

I see the room in which our bodies
have been alone with each other,
our bodies, saved out in space,
carried along by a quiet time.

CAMERA OBSCURA

I would like to stick my hand
through your big pupils,
groping, to look for
the images you save of me.

I would feel my own sight
and know where I touched you,
where in you I'm still warm
and where I've already cooled off.

interment

Unbelievable
how you can sit still:
your hands are in another country,
your eyes look past your death,
your smile is as softly as possible
turned to stone.

the sleeping beauty

I can go through walls
and at a wink from me
twilight and dawn flow
back into day and night.

I withdraw from myself
with the light rustle
of a birth which has not yet
thrown forward a shadow in time.

The child doesn't cry yet:
the doctor and the woman
rigid in union and expectation,
keep life and death in balance.

The placenta,
kneaded out of father-meat and child's blood,
is instruction, tenuous prescription
for what a boy does with a girl.

The scent which rises from the rose
lingers timelessly
and still,
the way the future adjusts itself to my will.

ars amandi

It is not true by a long way
that the two most beautiful people in the world
love each other the most,
the most or the best.

As it is not proved either
that the two ugliest people in the world
love each other the least,
the least or the worst.

The dark obeys
other rules than the light.

What our fingers feel
doesn't get disproved by our eyes.

He who says admiringly:
'Sweetheart, how ugly you are!'
says it rightly.

jews on the terrace

Laughing and well-dressed, the men dark,
the women fat, with hair made blond,
eating ice-cream, drinking coffee,
they look at each other.

The city has almost not changed on this square:
prosperity has replaced earlier poverty
and the arbitrary has made room for freedom
or, rather, free time.

Not a quarter of an hour from here where then was a city
in the city, old houses are being broken down,
ruins cleared away, in concrete and steel
the past being wiped out.

mobilization

It is not sufficient
to hate a general,
to wash one's hands in innocence,
or to march along in a parade
and get oneself beaten in the face.

Everybody lodges a grey, moustached ironeater
who is shamed by a sensitive wink
but after a simple order
sends his army of good-humored soldiers
rejoicing into death.

dictatorship

Speechless passers-by rest
themselves under the autumn trees.

There is inaudible shooting.

Spectators at the death
are prohibited.

Field-Glasses

Soldiers kneel in their own blood,
arise from the blood of others.

Armed surveyors survey the dawn:
the coming destruction is being calculated
and mapped carefully.

Gunners tune the silence
which is unbearable.

The enemy's distance gradually becomes
picturesque again.

Civilian

The war was a savage story
of others to which I listened
while the war raged on.

All the same I was tired every evening
as if I were lying in the front line
and when morning broke I
looked at the trees on the other side
of the canal as a sentry
looks out at no man's land
to discover that the enemy
had not yet moved himself in the light.
And when my sweat broke out at a bang
and I shrank in my armchair
I died for a moment.

intentions

I can have a leg amputated,
then limp home, after as much time as necessary,
like a soldier
who comes from the war.

I can go blind
because I've seen enough.

I can die with a curse
of happiness in my mouth.

I can live my childhood again,
have my parents rise from their graves.

I can pluck a feather from a sparrow
and remove
powder from the wing of a butterfly.

I can make a date with the sun
and have her come for nothing.

matrimony

I shall not strike you:
only not stroke you

I shall not bite you:
only not kiss you

I shall not overpower you:
only not play with you

I shall not kill you:
only not let you live

theology

Of course there is no god
as believers imagine him.

Don't let yourself be intimidated
or tempted into senseless discussion,
not even on television.
You have something better to do.

If god exists he may be
a computer of unimaginable circumference
by which all your lost moments
are being put on record:
so you can safely forget something.
But to talk about that
is senseless too.

God might just as well be a cent,
a zinc cent from 1943
(zinc because at that time they could
use copper for a better purpose
than for mere coin:
it takes capital to kill a single person
and it gets more expensive every year).

And certainly god is an old mill
always turning
to grind bread into meal,
prayers into remorse and resignation.

And therefore Jesus is not
the son of a god,
supposing that gods live
cat-and-mouse like us
and practice birth-control
(his resurrection is a fairy-tale for Easter,
an Easter-egg)
but he was a person like you,
in some things better or different than we.

I wouldn't like to pass my short life
with twelve apostles.
I wouldn't know what to do
with my nights,
let alone my mornings:
at the wash-basin, all that coughing
of grumpy men.

And further, god is a legend
on which the church
with all her popes and cardinals
has built in Gothic letters
the great joke of Christianity,
the church a mitred bunker in Rome
which at last must be blown up
with ordinary dynamite.

Be stalwart! Be neat!
Let your feet be washed by nobody
and certainly not by a priest
who when he smells sweat
immediately thinks of the last supper.

Go to bed with a free ritual
which you change to suit yourself
and according to who you're going to bed with.

Pray if you can't help it
but pray like somebody who is glad
that he is alone
and can walk away
or just roll over on his other side:
to be a Siamese twin is much worse.

There is no Holy Scripture:
language is started by people
in explanation of nothing
that is not wordplay.

Don't you find that sufficient?
It is already difficult enough to get any wiser
in Babel and at any other world convention
of theologists and philatelists:
life is a linguistic problem
and all theology
is poetry.

Words which spoken in silence
increase the silence.

Laughter:
yes, laughter above all.

God has died after never having lived:
it is his only miracle.

An old mill which turns forever
to grind bread into meal.
The most divine intoxication is being well-slept,
to see the world with clear eyes:
every awakening a good-bye,
a clear step
into a thank-god unclear future,
a miracle.

Born 57 years ago in Ijmuiden, Adriaan Morriën was evacuated to Amsterdam during the war, and has lived there since.

His first book of poems, *Heartbeat,* was published in 1933, followed by *Land Wind, Air-Raid Alarm, The Fatherland, Friendship for a Tree, Looking at the Clouds,* and *Poems of a Father.* A selection from these seven volumes, *Poems,* was was published in 1961.

Morriën's two most recent books of poems have been *Mothers and Sons* (1962) and *The Use of a Wall Mirror* (1968), from which the poems in this book are drawn.

He has also published five books of prose, both essays and short stories. He has edited several literary magazines in Holland, and is the only foreign member of the celebrated Group '47, which includes Günter Grass and Heinrich Böll.

For a number of years he has made a living as a literary translator, mainly from the French (Laclos, Balzac, Maupassant, Camus, Anouilh, Ayme), but also from the English German and Italian. His most recent work is *The Story of O.* In 1969 he was appointed lecturer at the Institutte for Interpreters and Translators of the University of Amsterdam.

He is married and has two daughters.

The Cover drawing is by Vapi.

Printed from Aldus linotype on Curtis Rag paper
in on edition of 1,000 copies by Don Gray at the
twowindows press in San Francisco.